"We live in a society exquisitely dependent on science and technology, in which hardly anyone knows anything about science and technology".

Carl Sagan.

Contents.

Foreword.

Never in the history of mankind has so much knowledge and information been so available to so many people, in so many ways, yet ignorance abounds.

This book is written with the intention of helping people to *notice.* It will not have huge lists of references as many of these will be given when evidence is presented making for easier reading.

The content of this book is not my opinion. It is based on scientific evidence, mainly by male scientists.

I have no axe to grind against men nor have I any religion to promote or put down.

This book is intended solely to stimulate thought and debate. Hopefully to save the shrinking Y chromosome and help women regain their rightful place in modern society, which is superiority, but we will settle for equality.

This quote by Sir William Golding is very accurate.

"I think women are foolish to pretend they are equal to men; they are far superior and always have been. Whatever you give a woman, she will make it greater. If you give her a sperm, she will give you a baby, If you give her a house she will give you a home. If you give her groceries, she'll give you a meal. If you give her a smile, she'll give you, her heart. She multiplies and enlarges what is given to her. So, if you give her any crap, be ready to receive a ton of shit."

Chapter 1

The Beginning.

To begin at the beginning. So, when was "The Beginning"?

According to the book of Genesis in the Bible, God first created man. This has justified treating women as inferior for centuries. Science has another version of "The Beginning." It's called "The Big Bang"!

This took place around fourteen billion years ago. If we look at the history of our planet as a calendar[1.] then the Big Bang happened in January 13.8 billion years ago. Billions of years before God created Adam.

The atmosphere consisted of noxious gases and a kind of mineral soup before life evolved.

If we start our journey on the Cosmic Calendar, then the Big Bang happened in January 13.8 billion years ago.

It was March 2 billion years later, before the first galaxies formed. The Sun and then our own planet Earth formed in September of the cosmic calendar some five billion years ago.

Conditions were very harsh. There was little or no atmosphere as we know it, there was no free oxygen.

The first forms of life appeared in October around 3.5 billion years ago.

These were a group known as Archaea. These first cells were extremophiles, and all were adept at surviving in the harsh early conditions.

Their names give clues to their habitats. Halophiles in extremely salt water, methanothermus found in very high temperatures, pyrodictium in porous walls of deep-sea vents and sulolobus in hot acidic sulphurous springs. Variants of these life forms can be found today in similar habitats.

Methanosarcina rumen is the methane producing organism found in cattle and sheep. Halococcus salifodinae thrives in the high salt concentration of the Dead Sea.

The first simple cells we have evidence for are called prokaryotes. They had no true nucleus and their genetic material was in single strands forming loops or coils within the cell. They had no individual organelles except ribosomes, which make proteins for all the cell's functions.

They were anaerobic, so did not need oxygen to survive and they had to find their own food.

By early November in the cosmic calendar 2.5 billion years ago, more complex single cells, eukaryotes, had evolved, some of which could make their own food via a process known as photosynthesis.

We humans, like all other animals, are made of trillions of eukaryotic cells.

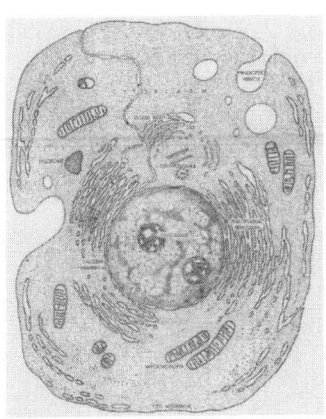

These eukaryotic cells had a true nucleus surrounded by a nuclear membrane, see diagram, and many other individual organelles, two of which were critical in evolution, mitochondria, and chloroplasts.

Mitochondria are the little sausage shaped, striped, organelles in the diagram. The DNA in mitochondria is different to the DNA in the cell nucleus showing that mitochondria must have at one time been separate organisms.

This could have come about if a mitochondrion was engulfed as food or was a parasite that could only survive in another organism and managed to survive and breed within the other cell thus becoming an organelle of that cell. This was such a crucial event in evolution that it was known as "The Fateful Encounter Hypothesis." We will return to this later.

Chloroplasts are the organelles in plant cells which carry out photosynthesis. Thus, these cells were autotrophs, they could make their own food, and oxygen was released into the atmosphere as a product of the process of photosynthesis.

These early plants were like beds of green floating algae, but they still carried out photosynthesis.

Then, by late November, or 1.8 billion years ago, there was free oxygen in the atmosphere. This was a *critical event in evolution*. Now that some cells could photosynthesize, oxygen levels began to rise and paved the way for more complex organisms to evolve.

By late November, 1.7 billion years ago, there is evidence of multi-cellular organisms.

From these original single celled Archaea all multi-cellular organisms evolved, including us!

But how did these simple multi-cellular organisms evolve into such complex organisms?

There are many theories as to how multi-cellular organisms evolved, but scientists lean towards these three. "The Cellurization Theory". "The Colonial Theory" and the "Symbiotic Theory." It is the last of these that scientists consider most likely and as such has been named "The Fateful Encounter Hypothesis."

Endosymbiosis is when two prokaryotes of different species live together. The smaller one living inside the cells of the host gains safety and wellbeing and eventually loses the ability to live independently. There is good evidence for this in several organelles, particularly mitochondria and ribosomes, which have their own DNA that is different from the DNA in the cell nucleus.

This evidence shows that both these organelles could have lived as independent organisms at one time.

Mitochondria are the cells powerhouse. They are found in the cytoplasm of eukaryotic cells and their job is to convert nutrients into energy for all the cells' functions.

They do this via Kerb's Cycle, also known as citric acid cycle, and store it in high energy bonds in Adenosine Triphosphate, ATP.

Mitochondria can occupy fifteen to fifty percent of the cell's volume. Without them we would not exist.

Biologists claim that these simple, single cells evolved into intermediate organisms, which then gave rise to amphibians, reptiles, mammals, including man. But these protists or "very first" eukaryotic cells were as simple as the common amoeba, so how did they reproduce?

Chapter 2

Reproduction.

These simple cells reproduced by binary fission. A process where the cell doubles its DNA and then splits into two genetically identical cells. This is simple asexual reproduction. No gender is assigned to the parent cell at this stage of evolution.

This kind of reproduction has many advantages. As it only involves one parent, no time or energy is wasted in looking for a mate. All the offspring are identical to the parent, and it is not a complex procedure, requires little energy, and animals can make use of favourable environmental conditions to produce lots of offspring.

Many organisms such as Komodo dragons, yeasts, whip-tail lizards, freshwater snails, plants, and fungi, still reproduce this way today.

The average size of these early cells was around 0.5 – 1 micron. How did they increase in size to the organisms we have today?

Three-dimensional growth i limited by surface area/volume ratio. As the size of a three-dimensional organism grows, its volume increases faster than its surface does, which causes problems for cells, so some started to group together.

A perfect example of this is the colonies of volvox, a beautiful free floating fresh water green algae, which form a colony known as a coenobium. A coenobium is an association of a fixed number of cells surrounded by a membrane. When the fixed number is reached, daughter cells break off to form their own coenobium. The word coenobium has an ecclesiastical origin where monks or nuns of the same order formed groups.

Volvox

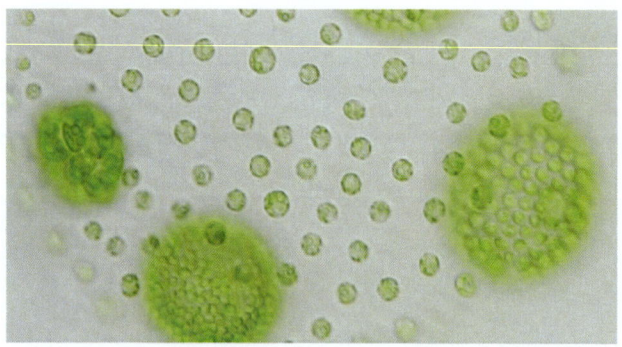

Another example of grouping comes from stromatolites. These are rock-like structures built by cyanobacteria, a blue green algae. These stromatolites still exist today. Some under deep water, but a large colony in Shark Bay Australia is exposed. Scientists have traced a micro-organism found in these rocks to one that lived 1.9 billion years ago. This is the furthest back any organism has been tracked using DNA tracing.

Stromatolites.

Up to this point in evolution we have only encountered organisms that reproduced asexually. Asexual reproduction does not involve the fusion of gametes which are haploid and carry only one copy of the chromosome.

It therefore produces offspring that are genetically identical to the parent and does not assign any gender to that parent.

This is simply cloning, and the parent passes on any mutations good or bad.

It also poses the question about gender. When, and more importantly why, did gender originate?

Chapter 3

The Existence of Men.

"All truth goes through three stages.

First it is ridiculed.

Then it is violently opposed.

Finally, it is accepted as self-evident."

Arthur Schopenhauer said this hundreds of years ago and it still applies today.

It's not yet clear which of Schopenhauer's stages we are at as regards the truth of how males came into existence. Probably the second but we are aiming for the third.

According to Professor Matthew Gage, University of East Anglia.[2]

"Almost all multi-cellular species on earth reproduce using sex, but its existence isn't easy to explain because sex carries big burdens, the most obvious of which is that only half of your offspring, daughters, will produce offspring. Why should any species waste all that effort on sons?"

Why would animals abandon asexual reproduction in favour of an inefficient and more costly sexual reproduction?

Sexual reproduction produces more genetic diversity within the species as the gametes combine, and like life itself, sex began in the oceans.

In 2014 a group of international scientists published a paper in the scientific journal Nature, and what they claimed was amazing.

Palaeontologist, Professor John Long of Flinders University said, "We have defined the very point in evolution where the origin of internal fertilisation in all animals began. It was in ancient lakes in a place that is now known as Scotland." They also traced it back to Estonia and China. We Scots have laid claim to almost every great invention that helped human progress, from the steam engine to penicillin. Now sex!

This group of scientists were studying a all fossil fish microbrachius dicki, and they concluded that this fish was the first to stop spawning and start copulating 385 million years ago! The scientists think this first attempt at copulation did not last for long and the fish reverted to spawning and it was a few more million years before copulation made a comeback in ancestors of rays and sharks.

However, one of the most interesting bits of information to come from this work was the evidence that there were both males and females around at this time in primitive species.

But it does not tell us how this came about. We must turn to another branch of science, genetics, to help explain how this came about

Chapter 4

A Mutated X.

All animals do not share X and Y chromosomes, but all mammals do.

Scientists at the Whitehead Institute[3] when comparing X and Y chromosomes made an amazing discovery. They found that some three hundred million years ago, our reptilian ancestors had an autosomal pair.

That is, both chromosomes were the same XX, which means they were all female.

But they also found what Dr. Steven Rozen called. "A Rotten X".

This was an X chromosome that had mutations.

One of these ancestral chromosomes had acquired a gene mutation for **maleness.** So, an **X** had become a **Y.**

A look at the Y suggests that in the process it had lost some genes. Perhaps this was the biblical "Adam's Rib".

More recent versions of maleness have been found by researchers working at Queensland University Australia and in Japan.

They were studying a gene on the Y chromosome that determines gender by producing an enzyme, Jmdl, in mice. Removing that gene produced **females** in mice that had the Y chromosome.

So far, the evidence shows that multi-cellular organisms appeared around 600 million years ago.

First maleness was detected in reptiles just 300 million years ago. This tells us multicellular organisms had been reproducing on their own for over 300 million years.

The scientific evidence currently tells us that these early species were all female, and these species have been evolving for some 300 million years more than the mutated X.

This poses the question. Have males evolved as far as females?

Comparison of anatomy tells us otherwise Many male bodies are still covered in hair as were our primate ancestors. Males have no womb and cannot produce offspring. Many also lose their hair at an early age.

This quote from a Gilbert & Sullivan opera describes it appropriately.

"Darwinian man, though well behaved, at best, is but a monkey shaved."

Many men no longer shave, and beards have become fashionable. So, one could add, that if in time he grows new tresses, evolution simply reverses!

Or to quote the late Stephen Hawking. "We only have to look at ourselves to see how intelligent life might develop into something we wouldn't want to meet."

You don't need to be a rocket scientist to see that Neanderthal man still exists.

A walk down the high street of any town will prove that

Chapter 5

The Genes.

Humans have twenty-three pairs of chromosomes. One of each pair is from the mother and the other from the father. All are roughly X shaped except for pair twenty-three which can be two X chromosomes if female, or one X and one Y for male.

Why a Y? Because it has lost some genes. What does that tell us about males?

Genes are the blueprint for making proteins that determine not only how we look but how we behave. Since all our other genes are X type males are obviously missing several genes on the Y.

How then did the earlier species reproduce

By parthenogenesis, also known as virgin birth.

This is a form of reproduction where the **female** gives birth to **females** and requires no input from males.

A myth-conception being Jesus. because had he been born of a virgin, he would have been *female!*

Yana traditions of Buddhism claim that there was "no carnal act involved" in Buddha's conception, but none of the original Buddhist texts claim this. Were this true, then Buddha too would have been *female!* Since females have two x chromosomes virgins can only produce females.

There are two kinds of parthenogenesis, obligate and facultative. As the name suggests obligate species can only reproduce that way and can only produce females.

On the other hand, species who practice facultative parthenogenesis can switch between parthenogenic and sexual reproduction and this, as we shall see later, still applies to human females.

So, when males are scarce, these species can still reproduce, but they will only produce females unless they are Komodo dragons.

Many case of facultative parthenogenesis have been reported on animals in captivity, such as the Komodo dragon Flora, at Chester Zoo and Sungai, a female Komodo at London Zoo, who produced four males from a clutch of twenty-two eggs although she had not mated with a male in two and a half years.

It is possible though, that some sperm from previous encounters had lurked around as it does in some species. For obvious reasons, no cases of parthenogenesis had been reported in the wild, until recently in Florida.

The population of the Smalltooth Sawfish, Pristis pectinata, a native of Florida rivers, has been declining over the last century and males were in very short supply. A group of scientists was assigned to monitor the population which had declined to 5%. They were monitoring the DNA of all new-born sawfish. The sawfish is no tiddler. It can grow up to four metres in length.

What they found was quite revealing. Said Doctor Kevin Feldheim,[4] Field Museum of Chicago, "Occasional parthenogenesis may be much more routine in wild animal populations than we ever thought."

Of one hundred and ninety fish that the group surveyed, seven had DNA that indicated that they had only one parent. These fish were exhibiting facultative parthenogenesis, proof that in extenuating circumstance females can and do reproduce without males. This is not exclusive to fish.

German-born, American biologist Jacques Loeb, over a hundred years ago, got chicken, rabbit and even monkey ova to fertilise using electrical stimulation and saline solutions.

He refuted the idea that spermatozoa were necessary for embryonic development.

His conclusion was.

"The male is not necessary for reproduction. A simple physio-chemical agent in the female is enough to bring it about."

More recent work done in Australia by Professor Peter Koopman at the Queensland Institute for Molecular Biology[5] has identified the gene on the Y chromosome that determines gender. He claimed that

"Most mammals, including humans and mice, are **programmed** to develop as **females** unless a specific gene on the Y chromosome called SRY, is present to trigger **male** development during embryonic life".

The human embryo is female up until about twelve weeks in the womb, even if it has a Y chromosome. Around nine weeks into gestation, genes on the Y chromosome must be activated to determine gender.

Who or what activates these genes?

The mother!

This means that should the mother be in poor health, stressed, or even lacking a nutrient in her diet, the process may not be fully completed.

This could account for many of the homosexual and inter-genders that are now coming to the fore in modern society and should send out the message that these states are not lifestyle choices, but developmental errors in the womb.

Chapter 6

Gender.

There are over 30 trillion cells in the human body. Each cell behaves like a mini human being. It breathes, eats, reproduces etc.

Mathematicians have calculated that at any one second there are thirty-seven billion, thousand, thousand, reactions going on in the human body.

That's thirty-seven with twenty-one zeros after it.

Anything we put into your bodies will affect some of these reactions.

What does that tell us about drugs? They are designed to tackle one thing but with so many reactions going on in the body they can affect many others.

With such a volume of reactions it would take forever for a drug to be tested properly so that it would not interfere with other crucial reactions in the body.

Ben Goldacre[6], in his book, "Bad Pharma" tells us how the pharmaceutical companies exist not to cure disease but to make profit.

In the late nineteen fifties and early sixties a drug was prescribed to many women for pregnancy related morning sickness. This drug, Thalidomide, caused untold harm.

Around ten thousand babies across the world were born with physical defects. Some had no arms, legs, or ears. Others had malformed internal organs. It was eventually taken off the market, but it took another sixty years before research told us exactly what was happening.

A study in the Lancet Medical Journal in 2019 found that pregnant women who took Acetaminophen, sold in the UK as

Paracetamol, a pain killer, doubled their risk of having autistic babies.

Many of our genes are like light switches and can be switched on and off. The cell proteins that help the genes switch on and off are called transcription factors.

Thalidomide acted by degrading transcription factors including SAL4.

This is what caused the devastating effects.

If a drug like this can cause so much harm to an infant in the womb, it is possible that other drugs, prescribed or otherwise, taken by pregnant women, for other causes, could easily interfere with the process of gender determination.

An article in the Journal of Urology claimed that Paracetamol can cause erectile dysfunction, so it is obviously altering the expression of genes that control hormones. This alone could affect embryonic development and prevent accurate gender expression.

Also, many other over the counter medications contain Paracetamol so anyone taking e.g., Lemsip Max for a cold or flu and Paracetamol for some other condition, could easily overdose.

On their website, Scotland's NHS Lothian, Liver Transplant Unit and Research Laboratory[7], claims that the commonest cause of liver failure is **Paracetamol poisoning.**

Acute liver failure

The Scottish Liver Transplant Unit continues to be the primary referral site for patients in Scotland with acute liver failure. Each year approximately 50 patients with acute liver injury are transferred to the Unit and the most common cause for this injury is Paracetamol poisoning. A large database has been developed and Dr Kenneth Simpson along with Dr Darren Craig have, over the past year, published a significant number of papers.

This is not people deliberately over-dosing but can come about if a person is on the drug for a long period of time or on several drugs that contain Acetaminophen.

Jacque Loeb used saline solutions in his experiments. These are salt solutions and for decades people have been told to lower their salt intake.

For years medics and government health institutes have been advocating a low salt diet.

Considering life began in the salty oceans this would suggest that salt is necessary for our wellbeing. The sodium/potassium pump is the neurotransmitter used by our brains to pass information across the synapse from one nerve cell to another.

It is not just drugs that can affect development. We are all made of minerals. A piece of flesh from a human when analysed will present about ninety minerals. A few of these are aliens that are picked up from pollution etc.

We should get all the minerals we need from our food, but that is not the case today.

As long ago as 1936 the American Senate published a document that claimed that "American soil is almost completely devoid of minerals." The following figures from the 1992 Earth Summit show the depletion of minerals from soil over the last 100 years.

North America – 85%

South America – 76%

Asia – 76%

Africa - 74%

Australia – 55%.

Europe – 72%

It would appear that Australia is the best place to live at present.

Pseudo food could be another reason. Look at the ingredients of any simple food product and you will see that it is laced with many chemicals. Some to extend shelf life and others merely fillers. Any of which could interfere with the baby in the womb.

One person who suffered greatly from gender determination was the brilliant Alan Turing.

- Source unknown.

Alan was the father of modern technology, pioneer of the computer, breaker of the

Enigma Code which enabled us to shorten and win the Second World War.

In his time men in the UK were imprisoned for being homosexual. In 1952 he was convicted of homosexuality and was put on hormone therapy.   He   was   threatened   with imprisonment   if   he   did   not   take   the hormones.

He took his medication, but it did him no good. In 1954 he injected an apple with cyanide and took his own life to avoid being imprisoned.

There is anecdotal evidence that Apple Computers took its name and Logo as a tribute to Alan. I hope they did, he deserves to be recognised for his brilliant work. Fortunately, in 2013, Alan was given a posthumous Royal Pardon by our late Queen Elizabeth,   and   his   unprecedented achievements were honoured.

In many countries across the world being lesbian, gay, bisexual, or trans-sexual is still a crime.

Currently in the USA the Governor of Texas has had a petition raised against him for targeting parents of trans-kids.

Florida's Governor, Ron DeSantis is currently trying to pass legislation referred to as, the "Don't Say Gay" Bill.

This Bill will prevent teachers from teaching students about gender issues and sexual orientation. Teachers who do teach gender issues could be fired or prosecuted.

People are being punished for a glitch in the womb. Babies with a Y chromosome may have all the physical parts, but unless the SRY gene is adequately activated, gender may be indeterminate.

So, does Y make the guy? No.

Gender activation is done by a physio-chemical stimulus in the womb. Should the mother be under any form of stress or on drugs, prescribed or otherwise, activation of SRY may not occur sufficiently to produce the correct gender.

Would we criminalise a baby born deaf, blind, or with Downs Syndrome?

Would we attribute these disabilities to lifestyle choices?

Why, with all the scientific knowledge we now have, is gender still attributed to lifestyle choice and criminalized in so many countries?

Chapter 7

Parthenogenesis.

Is Parthenogenesis still possible in human females?

It has been claimed that Zoroaster, Moses, and Joan of Arc were all virgin births but had this been the case then Moses and Zoroaster would have been female.

There is, however, some scientific evidence for parthenogenesis occurring beyond invertebrates and insects. But does the mechanism for parthenogenesis still lie dormant in women?

According to Dr Walter Timme, this vestigial organ still exists in women.

He gave a lecture in the 1930's to the New York Academy of Medicine entitled

"Immaculate Conception a Scientific Possibility".

There is a part of the uterus which. like our appendix, once performed a function, but no longer does in normal circumstances.

It can be found as remnants of glomeruli and tubules of the lower part of the mesonephros. It appears as scattered tubules between the uterus and the epoophoron.

Timme claimed that the parocarium of the female reproductive organs could spontaneously produce living sperm to fertilize the egg cell and create an embryo, making parthenogenesis possible.

Although he claimed that this organ was used to produce sperm, it is quite possible that it simply produced the chemicals needed for the physio-chemical stimulus that Jacques Loeb used in his experiments.

This vestigial organ had, like our appendix, been of use at some time in evolution and was now dormant, but able to perform when the need arose.

In 1955 Emmi Marie Jones gave birth to a daughter, Monica, and she had not had any contact with a man.

It took the medics some time to confirm that Monica had indeed been conceived via parthenogenesis.

Another case occurred in the 1970's in the USA. Laurie, who belonged to a sect called Breatharians, was a very tall woman. She was celibate and lived on raw food. At one point she fasted for a year and later gave birth to a healthy daughter by means of parthenogenesis.

Although it was not until the eighties that we had DNA testing, this case also was confirm

Chapter 8

Parasites.

The Y chromosome is not just a sex chromosome. Only a few genes are specific to male development. The rest are autosomal genes that are shared with the X chromosome.

Since the Y chromosome is a mutated X, are males mutant females that can only reproduce symbiotically in a host?

Or are they a different parasitic species?

Is this, like the mitochondrion, another "Fateful Encounter"?

We tend to think of parasites as organisms that invade and kill their host, but not all parasites do this. Parasitism is a type of symbiotic relationship between two species.

Usually, the parasite benefits at the expense of the host, but not always. Parasitism comes in many forms, and it shares two of these forms with parthenogenesis, obligate and facultative. As in parthenogenesis, obligate parasites are completely dependent on their hosts to complete their life cycle therefore they will usually not harm the host.

A perfect example of obligate parasitism is that of head lice. They die if removed from the scalp. Facultative parasites do not rely on their host to complete their life cycle and only perform as parasites, sometimes.

If you look at the behaviour of men, who must mate to reproduce their genes, it is similar to that of lower species.

Peacocks and many other animals have elaborate courtship displays.

Men start by dating, and once they have assessed their date, may proceed to courtship. Courtship usually leads to engagement and marriage.

Engagement and marriage are both marked by a ring. The ring is to ward off other men and let them know that this woman is spoken for.

Men do not usually wear an engagement ring, but more men are wearing wedding rings.

The engagement and wedding rings say, "she is mine". A typical parasitic expression.

On 26th January 2019 on UK's BBC Radio 4 programme "In our Time," one of the scientists on the team discussing parasites, Professor Steve Jones UCL said.

"In sexual reproduction it's the males who are the parasites. They are forcing their female hosts, as it were concubines, to copy their male genes."

Does that apply to all organisms that reproduce sexually, including humans? Yes. He did not, however, say that the males were the same species as the host they were invading

If we look at what happens when a human catches a parasitic germ, bacteria or virus, the immune system kicks in to kill the germ and get it out of the system.

This can lead to coughing, sickness, vomiting etc. What is one of the first symptoms of pregnancy?

Morning sickness. Called morning sickness, but it can last throughout the day and for months. This is the immune system trying to get rid of a parasite, which confirms that men are parasites.

On October 11th, 2016, on BBC' 4 "The Life Scientific" Professor Jim Al-Khalili, University of Surrey was interviewing Ian Wilmut, creator of Dolly the sheep.

At one point they discussed human cloning and Wilmut's response was

"We are replaced by an electric shock."

And who are the "we"? Men!

Flashback to Jacques Loeb! An ovum from any mammal can be fertilized by a physio-chemical stimulus to produce a female. Can the same be done with a sperm? No! Sperm need to escape to a host, as do all parasites, where their DNA can reproduce.

As was mentioned before, mitochondria, the cell's life force, have their own DNA which is completely different to the cell's DNA.

Males contribute no mitochondrial DNA to their offspring.

Since sperm must swim to their target their mitochondria are concentrated in the tail, and this drops off as the Acrosome on the sperm head, which contains enzymes, penetrates the egg. Sperm therefore contribute no "life force" to the egg they penetrate.

This has enabled scientists to trace the human race back to a handful of women in Africa, via mitochondrial DNA.

Chapter 9

Spermageddon.

Human female babies are born with all the eggs they will ever have.

This is estimated to be about two million

By puberty the number is thought to have reduced to about 400,000.

These begin to descend into the womb at puberty when the twenty-eight-day menstrual cycle begins.

Males also start making sperm at puberty. The sperm numbers are in the multi-millions.

The numbers start decreasing with age, but most men continue to produce sperm well into old age, although the sperm are not always healthy.

Diagram 1 shows a healthy sperm. With a long tail and well-shaped head. Diagram 2 shows an unhealthy one. The head is elongated and tail short.

1

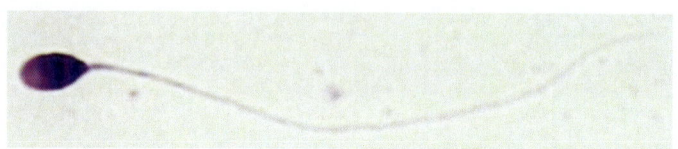

2

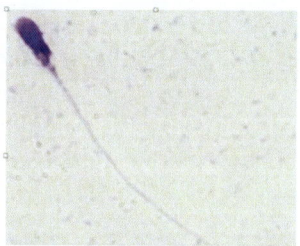

Source unknown.

Enzymes in the Acrosome at the head of the sperm are needed for the penetration of the ovum. If the head of the sperm is damaged these sperm will not be able to fertilize the ovum.

There is also conflicting research on numbers of genes for each gender.

The X chromosome has around 900 to 1,600 genes, some scientists say only 1,400. The Y has very few, around 200 but many of these are duplicates.

In February 2012, Matt Fearer[8] from Whitehead Institute published work that claimed that the theory of the "rotting Y" had been dealt a fatal blow.

However, in 2013, Professor Jenny Graves,[9] Head of Comparative Genomics Unit, La Trobe University Australia, published a paper that contradicted this.

(And now Maestro, a drumroll please!)

This leading Australian expert has predicted the **DEMISE OF MEN.!**

Because of the inherent fragility of the Y chromosome, it is shrinking away and will lead to the extinction of men in about 4.5 million years.

So, what! That's a long time away and we need not be worried. Or should we? If Professor Graves' team calculated the shrinking Y from the time of its first appearance millions of years ago, then we need to take account of modern living which is causing mayhem, with many diseases and conditions being directly related to the environment, the mess we are making of it, and to the poor quality of food we are consuming.

The Genomic Unit's findings were also splashed across the front page of U.K newspapers in 2013 yet little or nothing has been done about saving the Y.

Females have two X chromosomes so if anything goes wrong, they can help one another to fix the problem. Not so the Y chromosome. As it does not have another Y chromosome to interact with, it cannot repair damage and is shrinking.

According to the paper published by Australian Genomics Unit, the Y chromosome only has forty-five actively functioning genes. It has about two hundred genes but many of them are duplicates, and these are not counted as active genes.

Sperm are counted on a grid under a microscope, and it has been calculated that an ejaculation from a healthy male contains around 500 million sperm.

A huge study in 2017 by scientists from the Hebrew University, Jerusalem, conducted on 43,000 men from Europe, Australia, New Zealand, and North America found that sperm counts had fallen by 60% in the last 40 years

Andrology Australia reports that one in twenty men has a fertility problem.

However, it is not just the quantity of sperm that is important but the quality that matters most.

Sperm need good motility to reach their target and research now shows that many of them are not able to do so.

As in the diagrams on page 64, some lack a proper tail and in others the head shape is not normal suggesting that these sperm are unable to fertilize.

The Y chromosome is used as a sign of masculinity but only one gene, SRY determines gender. The Y contains very few other genes and is the only chromosome that is not necessary for life.

Women manage well without one.

That together with the shrinking Y should make us all sit up and take notice.

Chapter 10

Men Behaving Badly.

Males of all mammalian species have been behaving badly since the X chromosome became a Y.

As genes are responsible for everything in our lives, and as science has proved, males are missing many genes and cannot produce offspring themselves, then their genes will do anything to reproduce themselves.

Males of all mammalian species had to ensure that their genes were passed on or their species would become extinct. To do this they needed to dominate females

We need to go back a long way in time to see how women became subjugated by men.

Ancient Homo sapiens lived in tribes and clans. In evolutionary terms males simply spread their genes everywhere. They even inter-bred with other hominin groups like Neanderthals.

DNA tracing has shown that Homo Sapiens mated with another archaic human group, the Denisovans who existed around 30 to 50 thousand years ago.

Not only did men mate with other hominids, but incest was common. Marriage as we know it did not exist.

The first marriage recorded took place around 2,350BC and it became a popular institution in some cultures, mainly Greek and Roman.

These marriages were not based on love, but to "own" the woman, preserve the father's biological heir, and often to act as an alliance between families.

Leap forward to the sixteenth century. Henry the 8th. King of England. England, like most of

Europe, was Catholic at the time and divorce was not available.

Henry took the throne in 1509, when he was only seventeen.

Like most men, he wanted a male heir to succeed him.

His first wife was Catherine of Aragon, daughter of King Ferdinand and Queen Isabella of Spain and the widow of his older brother, Arthur. After multiple births, the only child to survive was a girl, Mary

Henry then took the law into his own hands. He asked the Pope to annul the marriage.

The Pope refused so Henry decided to leave the Catholic Church.

He then declared himself head of a new Church in England, the rules of which allowed divorce!

Wife number two was Anne Boleyn.

Anne was accused of adultery, but it is thought that she was beheaded because she did not produce a male heir.

He went on until he had taken six wives. He also had many mistresses, some of whom may have produced sons, but bastards were not acceptable as heirs.

It's a pity we did not have the science of genetics at that time. He could have been informed that women can only produce females, unless an egg is invaded by a Y sperm. Obviously, Henry's Y sperm could not swim as well as his X sperm.

There's an old rhyme that used to be quoted in British history classes. "Divorced, beheaded, died, divorced, beheaded, survived."

Anne Boleyn was only in her early thirties and his fifth wife Catherine Howard, in her twenties when they were beheaded.

This was the behaviour of a monarch who had control over a country and all the women in it.

Chapter 11

Modern Men Behaving Badly.

One would think that with history and education modern man would behave differently from his ancestors, but this is not so.

Sexist laws prevailed and it was not until 1929 that all women in the U.K. over the age of twenty-one were given the right to vote in a general election.

The U.K. continues to strive towards gender equality but there are still many areas where men and women do the same work, but the women are paid less than the men.

Up until 1982, it was legal to refuse a woman entry to a public house. These were regarded as "male environments".

It was 1990 before women were taxed independently from their husbands.

In 1922 Georgina Ballantine from Perthshire, a nurse, a registrar, and a keen salmon fisher, caught the largest salmon ever caught by rod in British waters.

Georgina and her father.

The picture is from a placard by Caputh bridge near where the fish was caught. The salmon weighed 64 pounds and was caught on the Glendelvine beat of the river Tay.

It was so heavy it took two and a half hours to land, and her father helped land it. A cast was made of this large fish and can be seen in a Perth Museum.

A water colour was also painted and hung in the Flyfisher's Club in London. Despite having caught the salmon, Georgina was refused entry to the Flyfisher's Club to see the painting because, she was a **woman.**

Ten years later the Imperial Japanese army between 1932 and 1945 provided brothels of "comfort women" for soldiers

These were women, and often young girls, who were dragged off the streets into cars and taken to these brothels.

Many did not survive the war years.

This article from "Inside History" describes another episode of male parasitism.

"On December 13, 1937, Japanese troops began a six-week-long massacre that essentially destroyed the Chinese city of Nanking.

Along the way Japanese troops raped between 20,000 and 80,000 Chinese women."

Had these armies consisted of women would they have needed "comfort men" I think not.

Even today rape is still used as a weapon of war.

This is a report from the Telegraph last year.

Women rush to hospitals for emergency treatment in Tigray region as rape is used as weapon of war

By Lucy Kassa
and Anna Pujol-Mazzini

It described how hundreds of women in Northern Ethiopia were going to hospital for emergency contraception and HIV protection after being raped by Eritrean and Ethiopian soldiers fighting in a civil war.

In one case, soldiers entered a house where a young woman was with her older brother. Her brother tried to protect her and was killed. She was then raped by several soldiers, next to her brother's corpse!

However, it is not only in war that rape happens.

In March 2021 a British police officer, Wayne Couzens, was jailed for life for the rape and murder of Sarah Everard. This man has a wife and children who will now suffer because of his behaviour. These children will now grow up without a father and probably be subject to verbal and other abuse because of their father's behaviour.

The police force is supposed to protect us from criminal activity, not perpetrate it.

Many women are now very wary of contact with police.

Chapter 12

Tradition.

The fact that women are the child bearers has contributed more to their subjugation than anything.

With men sowing their seed wherever and whenever they liked, women were constantly either pregnant or caring for babies. Human babies being helpless, compared to other species, women were tied to childbearing and rearing.

In places like Turkey hareems were common. Men had access to all the sex they wanted.

However, the poor servants or slaves who had to supervise the hareem were castrated and were known as eunuchs. This was done to prevent them from behaving like their masters.

The earliest records of intentional castration are from the city of Lagash in Sumeria in the second millennium BCE.

Later, religion stepped in. Marriage was introduced, probably to give men a constant outlet for their sperm. It had to be conducted by a priest. Women had to consent to obey their husband and birth control was unknown.

The woman had to take the man's name. This was to claim the woman as "his" but was probably a good thing as it would help prevent people who were closely related from marrying, and prevent incest, which can cause mayhem with gene mutations.

Men were given 'Conjugal Rights" and in the U.K. up until 20th century men could take their wives to court if they did not perform these rights, and the judge would order "Restitution of Conjugal Rights."

What rights did the women have? None!

Some religions still allow men to have more than one wife. Islam allows four wives and even the Mormons practised plural marriage up until the 20th century.

Despite the introduction of marriage, many men still behaved badly.

This can be explained by looking at the chromosomes.

X   Y

Chromosomes are strings of genes.

Genes are recipes for proteins that are responsible for everything in our minds and bodies. These proteins control not just how we look but how we behave and think. Some are fixed, as in eye colour, skin colour etc.

But most can be switched on or off. It is obvious just from the shape that the Y has fewer genes than the X

Doctor Bruce Lipton did some legendary work showing that our genes are controlled not just by the environment, but by our **perception** of the environment.

We are now into Epigenetics, where it is said that.

"Genes only load the gun; the environment pulls the trigger."

The environment being one's lifestyle

But it can only pull the trigger on the **genes that are still there.**

A look at some police statistics from the U.K. is enough to show us that the Y chromosome has lost some very important genes.

Particularly in the prison population figures.

In the year 2020 in the United Kingdom some 770,420 men were in prison and only 3,410 women!

The arrests were 15% female and 85% male. Convictions were only 27% female but 74% male

As a percentage, prison population was 5% female and 95% male.

United Kingdom statistics for rape reported to the Police in 2020 were 153,000. Convictions were only 1.400.

Men also occupy most of the prison cells across the world, not just in the U.K. Why is this?

There could be many reasons for males breaking the law, poverty, alcohol, drug use etc.

But the most likely one is that the shrinking Y has lost genes that control behaviour. Females still have those genes.

From the "Big Bang" until 300 million years ago all the evidence points to females being the first gender on the planet.

How then did females come to be regarded as inferior to males?

Some men intuitively know that women are the primary gender and try to cover by being manly, tough, strong, etc.

This quote from Professor Francis Lester Ward puts it very succinctly.

"Women are the race itself ------ the strong primary sex, and men the biological afterthought."

Men are very competitive. As are all parasites.

Not only in finding a mate, but in business and in games they invented.

We just have to look at the games men invented, and still play, to see where their instincts lie. Teams of men compete against one another.

What is a football match? A bunch of guys trying to get a ball into a goal.

For ball, read sperm, and for goal read vagina. The same applies to golf and other male dominated sports.

Men have been behaving badly since time began.

One of the earliest cases around 1,400 years ago was that of Hypatia (Hy-pay-sha). She was born around 350 -370 AD.

Her father was a Greek mathematician at the library in Alexandria, which was the seat of learning of its time.

People travelled the world to visit the library.

Hypatia was a philosopher, teacher, mathematician, and astronomer. Being an intelligent woman cost her her life.

She was dragged from her carriage in Alexandria by a fanatical Christian mob.

They took her to a church, stripped her naked. It is thought that they flayed her or beat her to death and then tore off her limbs before burning her remains.

Her crime? Being an intelligent woman who had chosen not to marry, but to remain working in the library.

One of the first women to protest for change in Britain was Emmeline Pankhurst. She organized the Suffragette movement and tried to get the right for women to vote. What did she get?

Frequent imprisonment!

Women did eventually get the vote, but she suffered mightily, and this was in a so called "civilized" country.

Women in other countries are not nearly as fortunate as Pankhurst.

Chapter 13

Culture.

Many cultures, particularly in the Middle East, to this day, treat women as chattels.

They are not allowed to go to school, are not allowed out unless accompanied by a man and can be bought and sold.

Some of these restrictions are also based in religions which are still based in the seventh century.

Two years ago, Care2 petitions asked people to sign a petition asking Google and Apple to take an App down from their sites.

It was an App that allowed Saudi men to spy on their wives. There was no App for women to spy on their husbands!

Carmen Bin Ladin[12], sister-in-law to Osama, in her book "The Veiled Kingdom" gives a unique insight into how women in Saudi Arabia are restricted and suffer. She managed to escape with her daughters and now lives in Switzerland.

Princess Latifa, daughter of the ruler of Dubai, tried to flee the country in 2018. She was trying to escape to America.

In a BBC video, Latifa claimed that she was dragged off the ship by commandos, drugged, and then flown back to Dubai.

She was locked up in a villa with guards outside and all windows locked. She had no contact with the outside world.

Dubai and the UAE say she is safe and being cared for by her family, but nothing has been seen of her for some time.

United Nations Human Rights Office were concerned for her safety. In 2021 they asked the UAE to present a proof that Princess Latifa is alive, nearly two months later, on 9th April

2021, the organisation said that while the Emirates stated that Latifa was being cared for by her family, the country failed to provide a "proof of life" for her.

Her father, Sheikh Mohammed bin Rashid Al Maktoum, whom she claims organised her capture, is a billionaire racehorse owner who comes to Britain to see his horses race at Ascot.

Two more things that have contributed to the subjugation of women, are culture and religion.

It was probably around the time when religions started that many of these restrictions were introduced.

Perhaps to protect women because many men could not control themselves.

Culture is a set of beliefs and behaviours created, usually by men, to navigate life.

There is an ocean of beliefs, ideas, rituals, good and bad, that rule our lives.

In fact, from birth onwards, we pick up these beliefs from parents, educators, religion etc. and most people continue to believe in them until they die.

One of the worst cultural practices that is still in force is FGM – female genital mutilation.

Young girls are subjected to horrendous genital mutilation, often without clinical instruments or anaesthetic. This is a barbaric act where scissors, razors and knifes are used to cut girls genitals. But worst of all when these tools are not available pieces of broken glass are used!

They are then sewn up with a needle and thread!

The reason given for this is that it should control the girls' sexuality so that they remain virgins until married.

The following figures are from UNICEF Global Database 2016 of estimated victims of FGM.

They list thirty-eight countries that practise this abhorrent mutilation.

You can also check out the following

Desert Flower website[12]

Country	Estitmated prevalence of female genital mutilation in girls and women aged between 15 to 49 years
Benin	9%
Burkina Faso	76%
Cameroon	1%
Central African Republic	24%
Chad	44%
Congo	5%
Côte d'Ivoire	38%
Democratic Republic of Congo	5%
Djibouti	93%
Egypt	87%
Eritrea	83%
Ethiopia	74%
Gambia	75%
Ghana	4%
Guinea	97%
Guinea- Bissau	45%
Kenya	21%
Liberia	50%
Mali	89%
Mauritania	69%
Niger	2%
Nigeria	25%
Senegal	25%
Sierra Leone	90%
Somalia	98%
Sudan	87%
Tanzania	15%
Togo	5%
Uganda	1%

Apparently, the current pandemic has escalated FGM as girls are not at school and teachers are not able to shield them. Girls as young as seven, are being mutilated. Are boys cut? No! The males can sow their seed where they please until marriage and after.

In Afghanistan a man is legally allowed to kill his wife if she does not bleed on their honeymoon. He may also cut her tongue out if a meal she has prepared displeases him.

He, on the other hand, is allowed to sow his seeds anywhere. Afghan men who have a guest will often offer them one of their wives for the night!

So much for culture and tradition

Chapter 14

Science and Religion.

The difference between scientific evidence and cultural evidence is that science, through experiment and tracing, gives us the truth with evidence as it stands now. It does not stand still.

Further evidence may be found to support or contradict existing evidence.

Not so with religious dogma!

Dogma is fixed and cannot be changed. This has worked to women's disadvantage for millennia.

I think this quote from Jon Stewart, an American, puts it in perspective.

"Yes, reason has been a part of organised religion ever since two nudists took dietary advice from a talking snake."

Religion plays a major part in the lives of most people in Arabia.

In February 2019 a six-year-old Saudi boy Zakaria, was beheaded by a taxi driver in front of his mother, just because he found out they were Shia.

The mother had sent blessings to Mohammad as she entered the car. The driver was Sunni, the prominent branch of Islam in Saudi Arabia.

If, as the evidence, tells us, all the early species that reproduced asexually were female, why have women been treated as inferior, second-class humans for millennia?

Some cultures claim that women need to be protected. From what?

Parasitic men! Biologically females have a cut-off date when they are no longer fertile.

This does not apply to males, most of whom continue to produce sperm, albeit diminishing, all their lives.

As men cannot reproduce parthenogenically these sperm need to escape to a female to reproduce their genes.

There's an old saying from mariners. "Men are like Camp Coffee. Aye ready!"

With hundreds of millions of micro-tadpoles in their pants all screaming "let me out" it's understandable that men are easily turned on.

Ah! I hear you say. What about prostitutes

They are doing all other women a favour! If some men cannot control themselves, they would be raping other women to release their load if the services of prostitutes were not available.

With evolution and our enlarged brains, the majority of men can and do control themselves, but not all men.

It was probably around the time when religions started that many of these restrictions were introduced.

Perhaps not to protect women, but to help men control themselves.!

Some cultures still have the practice of selling their children – girls, not boys – into marriage for money or goods, called dowry.

A case was reported in October 2018 where a 16-year-old girl's father AUCTIONED her on FACEBOOK!

This was in South Sudan where it is still "the culture" to sell child brides.

The horror of this case was one, that Facebook did not take it down quickly enough to prevent it; they took nearly two weeks!

And two, that five of the bidders were high ranking government officials!

The girl's father received 500 cows, 3 cars, and $10,000 in exchange for his daughter.

More evidence of the parasitic behaviour of men.

These officials could not find a partner through normal social means!

This in the 21st century! It beggars belief!

Plan International's Country Director in South Sudan, George Otim, said.

"That a girl could be sold for marriage on the world's biggest social networking site in this day and age is beyond belief."

Chapter 15

Tyranny Against Women.

We have moved a long way from Neanderthal man but few if any improvements have been made in many areas of the world.

As a member of Amnesty International I'm constantly reminded of the way some of these cultures treat women.

In a recent case in Iran two women, Yasaman and her mother Monireh were sentenced to sixteen years imprisonment.

The charge?

Inciting prostitution!

How did they do that? They simply said that women should be allowed to wear what they want, and it should not be compulsory to wear the burqa.

Fortunately, Amnesty International campaigned on their behalf and their sentence has been reduced to nine years and seven months.

Unfortunately, their friend, Mojgan, who was also arrested, must serve the full sixteen years.

These women are abused physically and sexually in prison and Amnesty is. fighting for their release.

If men can't control themselves if they see a woman who is not shrouded in a burqa, that is their problem, not the woman's!

The face veil was once part of some women's dress in the Byzantine Empire and was later taken into Muslim culture when the Arabs conquered the Middle East.

The traditional dress of Afghan women is colourful and does not cover the face.

Burqas are not the traditional garb of females in Afghanistan or any other country in the East.

Now that the Taliban have returned some Afghan women have started an online campaign called

"Do not touch my clothes."[13]

Now a ghost culture is being enforced in our country, demoting women to non existing beings - Mariam (@MariamBaraky)

The last phrase in the clip describes how women are regarded in many countries today.

Culture, like everything is of its time. It is just a set of rules which may have been useful in its time but much of it is no longer necessary.

Just because a practice may be cultural does not mean that we should not criticize it.

In the past it was 'cultural' to burn women at the stake in Britain if they voiced an opinion that was any different to the men, but we called them witches. Thankfully we have moved on from there and had a Queen for seventy years.

Alas many countries still treat culture as dogma, that which can never be changed!

Only in the year 2018, were women in Saudi Arabia allowed to drive cars. Not all women though, just certain categories. And to show just how powerful they are, the government imprisoned Nassima, the woman who started the protests to get women the right to drive.

They took her point and let some women drive but imprisoned her for starting the peaceful campaign. They not only imprisoned her, but also others who campaigned

Amnesty International reported that these women were beaten, tortured with electric shocks, flogged, and sexually abused whilst in detention.

 And who performed these cruel tortures? Men!

Worse still is the case of Nasrin Sotoudeh an Iranian human rights lawyer

She was sentenced in 2019 to 38 years and 148 lashes! Her crime?

Peacefully campaigning for women to have the right to choose what they wear!

Only a male dominated regime could hand out sentences like that!

Perhaps they also get pleasure out of being

"a dominant" and enjoy watching or delivering these awful lashings!

These letters from Nasrin to her children would bring tears to anyone's eyes.

To her daughter she wrote. "My daughter I hope you never think that I was not thinking of you or that it was my actions that deserved such punishment. Everything I have done is legal and within the framework of the law. It was then that you lovingly caressed my face with your small hands and replied".

"I know, Mummy. I know.

To her son she said.

"How do I tell you where I am when you are so innocent and too young to comprehend the meaning of words such as prison, arrest, sentence, trial and injustice." You asked me, "Mummy are you coming home with us today" and I was forced to respond in plain view of the security agents."

"My work is going to take a while, so I'll come home later."

Nasrin has been temporarily released and reunited with her husband and children because of the Covid 19 pandemic but Amnesty is fighting to prevent her return to prison.

As Nasrin said to her children, everything she did was within the law, but she was imprisoned and tortured. Now her husband Reza is facing six years imprisonment under the same charges. If he is imprisoned these children will be devastated.!

When Azza Soliman, a prominent lawyer in Egypt dedicated her life to defending victims of torture, arbitrary detention, rape, and domestic abuse, she was labeled a "spy" and a "threat to national security" by the authorities.

The only threat she posed was to male dominance.

She was put under surveillance, harassed by security forces on trumped up politically motivated charges, all for helping victims of abuse and poverty with legal aid, all of whom were women.

Worst of all was a case reported by Amnesty three years ago.

The following is straight from Amnesty's newsletter.

"Your eyes do not deceive you. An unelected all-male village council in India has ordered that 23-year-old Meenakshi Kumari and her 15-year-old sister are raped. The sentence was handed down as punishment after their brother eloped with a married woman. They also ordered for the sisters to be paraded naked with blackened faces. Nothing could justify this abhorrent punishment."

What does that say about men? One young man breaks the rules, and his sisters are punished!

It is also probable that the woman the brother ran off with had been sold as a child bride to a much older man, and one can't blame her for running off with someone her own age, whom she obviously cared for.

Nothing could justify this barbaric punishment of innocent women.

In Folkstone, UK. A "grooming gang" of Ethiopians is currently on trial for raping a girl from the age of twelve. There are ten men on trial, and they have a mixture of Islamic and Christian names.

There was an article in The Daily Mirror about a survivor from abuse, Joanne Phillips from Telford. It describes an incident where a man who had abused her for four years held a crossbow to her stomach when she told him she was pregnant. Joanne has published a book about how she was trafficked around the country and raped by hundreds of men from the age of twelve.

Many of these men from other countries regard white women as "easy meat."

This reminds me of a horrific incident experienced by a friend. She was living in Iran as her husband was working there.

One day she hailed a taxi to do some shopping. She realized that the driver was going in the wrong direction, and she commented on this.

The driver told her to shut up as he was taking her to a brothel where all white women belong.

Fortunately, at a traffic jam, she was able to open the door and throw herself out onto the road where she was rescued.

This took place a few decades ago but the trauma has left its mark on her to this day.

Violence against women in Turkey is rampant. President Erdogan has withdrawn from one of the most important international treaties in history, to prevent violence against women.

In 2020 three hundred women in Turkey were murdered, the majority by their partners.

These are just a few examples of what women still face in many parts of the world today. What can be done to change the minds of the male dominated regimes that still abound across the world?

Chapter 16

An Androcentric World.

Since the mutated X appeared on the planet, its genes had to reproduce to survive. To do this, males of all species had to be in charge of everything

Dictators and tyrants have been around for centuries, and all were male.

Since men only have forty-five active genes on the Y chromosome it is easy to see why dictators and tyrants behaved the way they did.

In our lifetime, Hitler was one of the first dictators. He tried to wipe out the Jews under the guise of "Ethnic Cleansing".

Stalin in 1932/33 was responsible for the Holodomor, a famine that killed around 3.9 million people in Ukraine.

This famine did not happen because of climate or any other catastrophe. It was brought about by a dictator who wanted to replace Ukrainian small farms with state run collectives

Why? He wanted to punish the Ukrainians for their independent minds as he thought this posed a threat to his totalitarian regime.

At one point after the second world war many countries were ruled by dictators.

Iraq was ruled by Saddam Hussein, China by Mao Zedong, Syria by Bashar al-Assad.

These are just a few of the tyrannical dictatorships where women were treated like prisoners.

They had to do what they were told by men or suffer.

Men in these countries regarded women as being put on the planet solely to supply the needs of men.

Females were not put on this planet to serve men. They were on the planet long before the mutant X appeared in reptiles.

We are now into the 21st. century and Ukraine has been invaded again by Russia. This time by Vladimir Putin.

It is terrifying to think that men who have only forty-five active genes on the Y chromosome have the power to press the nuclear button!

Chapter 17

A Look at the Missing Genes.

It would appear from the prison statistics that many men have lost the genes for controlling their behaviour.

But as genes do not always work individually, it might be difficult to find out which ones are responsible for such behaviours.

Other contributing factors to male behaviour are Asperger's syndrome and autism. One in five males is on the Asperger's /autism spectrum and this could be a result of missing genes. This does not mean that they have all been diagnosed but comes from meta statistics.

This together with the increasing number of boys in school who are now diagnosed with

ADHD, autism etc. generally called SEN, Special Educational Needs.

This could mean that the Y chromosome is shrinking faster than predicted by scientists.

Special needs pupil numbers have risen dramatically since the nineties and boys highly outnumber girls who are diagnosed as SEN.

There are many things that could account for this. Some scientists say that the food we are now eating has had a high impact on our mental wellbeing. But if this were the case for SEN then there would be as many girls on the spectrum as boys.

The spectrum of autism is very wide and covers a range of conditions that affect social interactions, communication, and behaviour.

Many famous people are said to be on the autism scale, and it is thought that people like Beethoven and Chopin were schizophrenic, and Mozart had Asperger's syndrome. Hitler was autistic and probably many other dictators, including the one who ordered the invasion of Ukraine recently.

Chris Packham, a BBC presenter was diagnosed with Asperger's in his forties but does a great job presenting Autumn watch etc. and it is not obvious that he has Asperger's.

One of the many symptoms of autism is an inability to communicate easily with others. This is very apparent in children with autism and makes life difficult, particularly in school.

A more lighthearted look at missing male genes would look like this.

The gene for multi-tasking. The gene for switching off lights.

The gene for lowering toilet seats! As a science teacher, when teaching the science of particles and how they spread I used these pictures taken from the school toilets.

Women's toilet seat with brown stains.

Men's toilet seat. No stains.

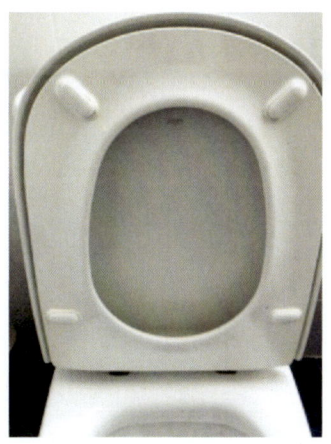

A look at these pictures will show how that which is not caught on the toilet seat, simply spreads into the bathroom for a distance of up to six feet. This spray from the open toilet lands on toothbrush, drinking glass, face cloth and towels. What is caught on the female seat that was closed? Shit!

The gene for reading full instructions before assembling things.

The gene for closing cupboard doors. The gene for putting things away – not down!

Obviously, there are no specific genes for these activities, but they are all part of male behaviour, and it may be that the shrinking Y has lost the ability to do many things.

Chapter 18

Shrinking Y.

The Y chromosome is facing extinction. All the current scientific evidence tells us that men are parasites, and the Y chromosome is shrinking faster than was predicted by the Australian Genomics Team.

But, as was said before, science is not static, and scientists could come up with more evidence proving that the rate at which the Y chromosome is shrinking is still increasing.

However, the most important gene on the Y that determines male gender is the SRY gene.

It is possible that this gene could disappear before others, and we could have an apocalyptic extinction of men much earlier than predicted.

A look at birth statistic genders shows how the Y chromosome is struggling to keep up with the X.

What are we going to do to save the Shrinking Y?

Do we want to see the demise of men? No. Life without men would be rather dull.

If science can help improve the state of those who can't control themselves, then it is time we had more **State Funded Research** around the world, concentrated on the shrinking Y chromosome.

We could also monitor the Y chromosome.

All men who are imprisoned, should have their DNA tested.

This would tell us if those convicted of rape and sexual offences have the same genes missing. They could be helped with gene replacement therapy. Some genes on the Y chromosome are essential for male survival.

We now have the science and technology to find out what is happening on the Y chromosome.

Change is necessary if we are to improve the lot of women throughout the world. Unfortunately change can be a subconscious threat.

Particularly to men in power.

Every newborn baby should have a DNA sample taken.

Analysis of this could tell us two very important things.

Are all males missing the same genes or does it vary? DNA could be tested as young boys grew up to see if the Y was constantly shrinking.

Are some women reproducing by parthenogenesis even though they are in a sexual relationship?

The DNA could also be stored so that criminals could be traced.

Does any male dominated government have the courage to implement this?

Perhaps when we have more countries where females are the rulers, we may achieve these things.

In the Periodic Table of the. Elements, Fe represents iron, so fe-male could be interpreted as Iron Man.

This quote, attributed to Mother Theresa, puts women into perspective.

W – Wheel of the family.

O – Ocean of knowledge.

M – Mirror of children.

A. – Address of love.

N – Navigator of the life boat.

To ensure that women are treated with the respect they deserve in all parts of the world we need.

M – Men who are aware that they are parasites and need women to reproduce their genes and will now treat women with greater respect.

E – Eager to Save the Shrinking Y

N – Never give up until they do both.

At the Vancouver Peace Summit in 2009 the Dalai Lama said something that went viral round the globe.

He claimed that he is a feminist and quoted.

"The world will be saved by the Western Woman."

Let's hope it is

References.

1.

Cosmic calendar on Wikipedia.

2.

https://researchportal.uea.ac.uk/en/persons/matthew-gage

3.

Whitehead InstituteMass.

https://wi.mit.edu

4.

Kevin Feldheim

https://www.fieldmuseum.org/about/staff/profile/551

5.

Queensland Institute for Molecular Biology
https://imb.uq.edu.au

6.

Ben Goldacre "Bad Pharma". Published by Fourth Estate.

7. NHS Lothian, Liver Transplant & Research Unit

8. Matt Fearer

https://www.labmanager.com/author/matt-fearer-whitehead-institute-for-biomedical-research

9.

Professor Jenny Graves. Head of Genomics Unit, La Trobe University, Australia.

10.

The veiled kingdom. By Carmen Bin Ladin. Virago Press 2004.

11.

Desert Flower Website

https://www.desertflowerfoundation.org/en/home.html

12.

"Do Not Touch My Clothes."

https://www.youtube.com/watch?v=51WEVXLCtMM

History Stories: https://www.history.com/news